Learn Statistics NOW!

Statistics for the Person Who Has Never Understood Math!

Minute Help Guides

Minute Help Press

www.minutehelp.com

Table of Contents

About Minute Help

Minute Help Press is building a library of books for people with only minutes to spare. Follow @minutehelp on Twitter to receive the latest information about free and paid publications from Minute Help Press, or visit minutehelpguides.com.

Introduction

Statistics . . . just the word sounds difficult, with visions of millions of numbers, percentages and Greek letters dancing in your head. But actually, we use statistics every day. The concepts are quite simple, and the math is often simple, too.

What is statistics? In the dictionary it is defined as the study of the collection, organization, analysis and interpretation of data. The first goal in statistics is to make descriptions that can be easily understood out of a huge mass of data. The second goal is to use those descriptions to make predictions. To statisticians, the accuracy of those descriptions and predictions are very important, and error calculations comprise a large part of statistics.

Probability

Statistics is often divided into two sections: descriptive (information about the past or present) and inferential (predictions about the future). This section will address the descriptive part, specifically probability. Inferential statistics will be covered in the next two sections.

Probability is the likelihood of something happening; it is just the way you use the word "probably" in everyday life. If you say, "I will probably be late for work today", you base that thought on some data, such as the time that you leave home. Maybe you are leaving for work later than usual. You compare what is happening now to similar data. For example, you think, "Most of the time when I leave this late, I do not make it to work on time." You make an estimate of what will happen. This is exactly what statistics is all about—describing what has happened in the past and using it to make an inference or prediction of the future.

Empirical Probability Formula:

$$P(E) = \frac{\# \text{ times } E \text{ occurs}}{\text{total } \# \text{ of observations}}$$

This is just the formal way of stating the way we automatically calculate a probability. $P(E)$ is the formal mathematical notation: P stands for probability and (E) stands for what the probability is of. The E is in parenthesis to indicate "of" and it is an E to signify that it is an event. So the left-hand side is simply the probability that event E will happen. The right side of the equation is just a fraction. The numerator (the top part of the fraction) is the number of times you saw E happen, and the denominator (the bottom part) is the total times you looked (called the "sample size").

So for the probability of being late for work, you would put in the numerator the number of times you were late, and you put the number of times you left this late in the denominator. Let's say you've left this late ten times and arrived late nine times.

$P(E) = \frac{9}{10}$ —when you have left this late, you have arrived late $\frac{9}{10}$ or 90% of the time.

Probability is usually expressed in a percentage; for example, we say that the chance of rain today is 50%. The statistic 50% is the same as saying a 50-50 chance or .5 probability. A 33% chance means you would expect to get that result 1 in three times. In everyday language, we say that 50% is an "even chance" and 33% is maybe, while 5% is unlikely.

However, something that simple is not necessarily accurate or a good predictor. This section will discuss the ways in which statistics, even simple statistics, can have errors. Errors are very important in statistics; when the statistics are wrong, then the decisions made based on them will be wrong, too. Perhaps it is just that you get to work earlier than you need to everyday or that you take an umbrella everyday. Or perhaps you print a newspaper that states that Dewey won the presidential election instead of Truman (which happened in 1948). Or perhaps you spend money on something unnecessary. When the spender is the government, it can have very serious effects.

"Garbage in, garbage out" is a catchy phrase coined in the 1960s to describe how you can input a lot of wrong or even nonsense data (garbage in) and get a result out, but this result is then wrong (garbage out). The complete phrase is "If you have garbage going in, then you will always have garbage coming out." Although it was coined by a programming instructor to apply to computers, it is equally applicable

to humans, describing failures in decision-making due to faulty, incomplete, or imprecise data.

This doesn't just apply to statistics; it is exactly what statistics is all about. Statisticians analyze how good the data is going in as well as how good the predictive value of that number is. In other words, a lot of statistics is concerned with how to make sure there is no garbage going in or coming out.

Probability is often simple to calculate, so it will usually not have any errors in its calculation. However, it can have other errors that are not as easy to spot, based on the data going into the equation. There are several terms that statisticians use about how to make sure their descriptive statistics are accurate and not garbage.

Sample Size

Let's deal with **sample size** first, because it is so important in every statistic. In fact, statisticians have a "Law of Large Numbers" that states that accuracy of predictions improves with more information. Another way of putting this is that the larger the amount of data (also called information, observations, or measurements), the more it will tend toward the actual (real or correct) average of that information. Better data in always means better descriptions out.

Naturally, the minimum sample size depends on each type of statistic and each type of prediction to be made. We will discuss that further in each section. The point here is to emphasize that sample size is so important to statisticians that they made it into a law.

A sample is any little piece of any bigger piece—like a sample taste when you go to Baskin Robbins. So sample size is just the amount of data going in, which can also be called the number of observations or data points. When you think you might be late, is it based on just one prior instance or lots of times? Size matters, a lot.

Example: The standard example is a coin toss, because it is intuitively obvious that the probability of tossing a heads is exactly 50% and the probability of tails is 50%. For flipping coins, P (E) = ½.

But let's say we didn't know this, or we wanted to test out the formula. So we pull a coin out of our pocket and we begin an experiment in flipping coins, recording what results we find—this is our data, and the number of times we flip the coin is our sample size. Say these are our results from 12 flips:

H H T H T T H H T H H T

(where H = heads, and T = a tails toss)

If you were to examine the probability of heads after four flips, you would have:

3 heads and 1 tails.

You would then calculate the probability of heads like this: 3/(3+1) = ¾ = 75%

After eight flips, you would have 5 heads and 3 tails,
and the probability of heads is   5/(5+3) = 5/8 = 63%

After twelve flips, you would have 7 heads and 5 tails,
and the probability of heads is   7/(7+5) = 7/12 = 58%

The point of this example is two-fold. First, a sample may not be an accurate measurement; we know that the real probability is 50%, and this experiment did not give us this number. Second, when it is inaccurate, it gets better with more data (observations, or in this case flips). Each of the above examples has exactly one extra heads and one too few tails, and it shows how the probability estimate coming out gets closer to what we know is true (50%) as the sample size increases.

A sufficient sample size depends on a lot of factors:

- What kind of statistic you are measuring,

- How frequently what you are measuring occurs (if it only happens 1% of the time you need a lot more data than when it happens 50% of the time), and

- The confidence level you want (how much error you are willing to accept in the estimate).

There simply is no hard and fast answer. Under 10 can be sufficient, and 1000 can be too few. The point here is just to describe what sample size is and why it is important. When the sample size is too small, it is an instance of "garbage in." You can calculate a number, such as an average, on a small sample, and that number will be accurate in terms of being calculated correctly, but it will be inaccurate in its representation of the whole—your small sample is not like the whole.

Sampling Error and Randomness

Sampling error refers to any of the ways that the sample you are using does not match the whole. These errors are not solved by simply getting a larger sample; even really large samples can have some kind of bias that makes them not representative of the whole. These errors usually happen because the selection of the sample was not designed correctly in order to get random input data.

> **Example:** Let's consider predicting the winner of a presidential election. You could survey an entire state with millions of people, such as California, the most populous state. However, that does not mean the winner in California will be the winner in the whole country. Even though you sampled 37 million people, and 37,000,000 clearly is a very large sample size, this is a sampling error because the sample was not randomly gathered—in other words, California is not representative of the country as a whole.

> Making a good prediction of election results does not require more input—37 million is actually much larger than what is needed. To make a good prediction, other samples are needed from the other states. The correct method would be a much smaller sample (such as 250) but within each of the 50 states, and then to predict each state individually. The correct method uses a sample size of 12,500 instead of the incorrect 37 million.

This is where the concept of randomness comes in. If you randomly pick some data throughout your whole population, that should take care of it, right? But this can be more difficult than it seems. You must avoid having the selection process of your sample affect the results in some unforeseen way.

> **Example:** It used to be that one way to get a random selection method for election results was to include in the sample only those people who had home phone numbers. The sample was otherwise carefully balanced for age, state, male/female, city/country, etc., and having a land line was a criterion only because of ease of obtaining the information. It worked for a long time, but recently it was discovered that the predictions were becoming less reliable, because younger people tended to have only cell phones. The random selection was, in effect, excluding an entire part of the whole and was no longer representative. Thus, the election prediction companies had to change their sample selection process to avoid this sampling error.

Combinations and Permutations

A related concept to probability is **combinations** and **permutations**. A combination is typically called an unordered set, while a permutation is an ordered set. For instance, if you order a pizza and a coke at a restaurant, it doesn't matter if you say "pizza and coke" or "coke and pizza." Both statements will achieve your goal, so here the order did not matter and it is a combination.

When the order does matter, it is a permutation. For instance, consider a combination lock. When you enter your four-digit code, the order in which you enter the numbers matters. If the correct combination is 4321, entering 1234 will not do the job. (Perhaps it should be called a *permutation* lock.)

Example: The Coke vs. Pepsi challenge. Let's say your friend claims she loves Coke and hates Pepsi and can always tell the difference. So you put her to the test by lining up just eight glasses, four with Coke and four with Pepsi. Her claim is that she can correctly identify the four Cokes. She will taste all eight and identify the four she is "certain" are Cokes. Is this a combination problem or a permutation problem?

The answer is combination. The order of her Coke picks doesn't matter; it just matters that she can pick out the Cokes. Let's look at the possible combinations. (The set of possible combinations is also called a **probability distribution**.) In this table, c represents a correct pick and w is a wrong pick, and each of the four glasses has Coke (C) or Pepsi (P). So there are four Pepsis identified in our list as P1, P2, P3 and P4, and there are four Cokes identified as C1, C2, C3 and C4. The order in which she picks them does not matter. (For example, she can pick C3, C2, C4 and C1, for all we care.)

cccc (her claim)	1 (there is only one way she can pick the 4 Cokes: C1C2C3C4)
cccw	16 (there are 4 glasses of each so there are 16 ways that she can pick 3 Cokes and 1 Pepsi in any order—
	C1C2C3P1, C1C2C3P2, C1C2C3P3, C1C2C3P4;
	C1C2C4P1, C1C2C4P2, C1C2C4P3, C1C2C4P4;
	C1C3C4P1, C1C3C4P2, C1C3C4P3, C1C3C4P4;
	C2C3C4P1, C2C3C4P2, C2C3C4P3, C2C3C4P4)
ccww	36 (possibilities)
	C1C2P1P2, C1C2P1P3, C1C2P1P4, C1C2P2P3, C1C2P2P4, C1C2P3P4;
	C1C3...(6 times);
	C1C4...(6 times);
	C2C3...(6 times);
	C2C4...(6 times);
	C3C4...(6 times)
cwww	16 (just the reverse of cccw above)
wwww	1 (P1P2P3P4)

We can calculate the probabilities as follows:

Probability of getting them all right:	1/(1+16+36+16+1)=1/70=1%
With one mistake:	16/70=23%
With two mistakes:	36/70=51%
With three mistakes:	16/70=23%
With all four wrong:	1/70=1%
If we add up the five percentages, we get:	99%

(It would be 100%, but we rounded the percentages off; the percentage for all right and all wrong is actually 1.4%.)

One really interesting point about this example is sample size. She is picking out only four glasses, but since the probability that she could do it by chance alone is so small, 1.4%, we will believe her if she picks the four Cokes. But even one mistake makes us feel that she could just have been lucky. Thus, with even just four observations here, we can have a high degree of confidence. This underscores how much the required sample size can vary depending on the statistic.

It is really tedious to list all those combinations! Don't you wish there were an easy way to calculate it instead? We're in luck—there is! That's where **factorials** come in. They may look a little confusing at first, but they save a *lot* of work and potential errors in making lists like the above.

Combination Formula:

$$\frac{n!}{r!(n-r)!}$$

It's easier than it looks. The factorial sign (!) describes a special calculation. Whatever number you apply it to, it means you multiply that number by each smaller number.

$4! = 4 \times 3 \times 2 \times 1 = 24$

$8! = 8 \times 7 \times 6 \times 5 \times 4 \times 3 \times 2 \times 1 = 40{,}320$

You can see that the numbers get very big very quickly, so as difficult as this formula may look at first, it is definitely worth not trying to map out all the combinations in a table!

In the formula, n is the number of things we are choosing from, and r is the number of things we are choosing. So in our Coke vs. Pepsi challenge, we would have do this:

$$\frac{8!}{(4! \times (8-4)!)}$$

That expands all the way out to: $\dfrac{8 \times 7 \times 6 \times 5 \times 4 \times 3 \times 2 \times 1}{(4 \times 3 \times 2 \times 1) \times (4 \times 3 \times 2 \times 1)}$ (because the (8-4)! is another 4!).

Quick, before you calculate all the way out, see if you can simplify the calculation.

You have 4x3x2x1 once in the numerator and twice in the denominator, so go ahead and simplify right away to

this: $\dfrac{8 \times 7 \times 6 \times 5 \times \cancel{4 \times 3 \times 2 \times 1}}{(4 \times 3 \times 2 \times 1) \times (\cancel{4 \times 3 \times 2 \times 1})}$

which equals this: $\dfrac{8 \times 7 \times 6 \times 5}{4 \times 3 \times 2 \times 1}$

Then simplify again: $\dfrac{\cancel{8} \times 7 \times 6 \times 5}{\cancel{4} \times 3 \times \cancel{2} \times \cancel{1}}$

which equals this: $\dfrac{7 \times 6 \times 5}{3}$

Simplify one more time: $\dfrac{7 \times \overset{2}{\cancel{6}} \times 5}{\cancel{3}}$

which equals $7 \times 2 \times 5 = 70$.

See how that matches the "all four right" and "all four wrong" possibilities above? Although you do end up with long strings of numbers, it is common to have most of them cancel out.

When order does matter, the numbers escalate even faster. Now 1234 is different from 4321. So the formula will save a lot more work in trying to figure out all the possibilities by hand in a table, and actually, the formula is simpler:

Permutation Formula:

$$\frac{n!}{(n-r)!}$$

So if we cared in what order our friend picked out the Cokes, it would work out to

$$\frac{8 \times 7 \times 6 \times 5 \times 4 \times 3 \times 2 \times 1}{4 \times 3 \times 2 \times 1}$$

which simplifies to **8 x 7 x 6 x 5 = 1680**

You surely would not have wanted to draw up a table with all of those.

Let's quickly calculate how probable it is to get that combination lock correct on one chance. There are 10 numbers (0-9) and 4 to pick (assuming a four-digit code), so that is

$$\frac{10!}{6!}$$

which equals $\frac{10 \times 9 \times 8 \times 7 \times 6 \times 5 \times 4 \times 3 \times 2 \times 1}{6 \times 5 \times 4 \times 3 \times 2 \times 1}$

which equals **10 x 9 x 8 x 7 = 5,040**

No burglar will get rich quick trying randomly to get lucky and figure out a four-digit code! That should make you feel a lot more comfortable about all security issues—your lock on your locker, your security code to your home alarm system, and your passwords on your internet accounts. And, on the internet, instead of just 10 digits, there are also 26 lower-case letters, plus 26 upper-case letters (if the password is case-sensitive) and special symbols. So imagine how unlikely it is for anyone to randomly figure out a password online. And they don't even know how long it is.

Probability summary

Probability can be thought of as time-related. It is a calculation of what percentage of the time something occurs. Usually that one number is based on past results and is then used as a prediction of the likelihood of something happening in the future.

If the information on past results is erroneous, then the prediction will be too. Statistics involves the study of the collection and organization of data to make sure that it is a random sample and of sufficient size, to try to make sure that the sample will match the whole—in other words, making sure the input data is accurate. Otherwise, the probability statistic will not be accurate, and any conclusions drawn from the data will not be accurate.

We all use these statistics every day, from predictions about the weather, to who will win the election, to whether we will be on time for work. We also hear and see them constantly on the internet and television. It is important to understand what kind of poor data collection techniques can make those probability statistics questionable.

Inferential Statistics

With inferential statistics, you are trying to reach conclusions from partial evidence. For instance, you want to infer from a sample that the sample probability can be applied to a larger population. Or you might want to judge whether differences seen between groups happen by chance or not. Or maybe you want to predict something in the future.

In other words, you are making an inference that extends beyond the immediate data alone, perhaps to a different time or a different sample for which you do not have data.

A large part of statistics is involved in evaluating how correct the inference or forecast is expected to be as well as how reliably you can apply that inference from the sample to the larger population. That's called the **level of confidence**. In testing, the level of confidence is so important that it is required for it to be specified in advance.

This section will look at some of the more basic types of inferences made from statistical data, plus some important inference terms, examining a single variable such a stock prices. The next section will examine more complex models with multiple variables.

Average and Arithmetic Mean

One key statistic used everywhere is an average. It is so commonly used that we don't even think of it as a statistic. It is popular because it is easy to calculate, and it can boil down even a huge list of numbers to a single value, so it makes a mass of numbers understandable. Everyone knows how to calculate an average, which is more accurately called an "arithmetic mean". (That's only because there are actually different types of averages, but the arithmetic mean type of average is the one commonly meant.)

Arithmetic Mean Formula:

$$A = \frac{\sum\limits_{i=1}^{n} x_i}{n}$$

Once you get past the symbols, this is a calculation you make every day!

A just stands for the **arithmetic mean**, which is the more exact mathematical term for average.

The funny-looking Σ is an uppercase S for the Greek letter **sigma**. It is a mathematical symbol meaning **sum**, or **add up**. What you are summing goes to the right of the sigma—**x**.

x is just the normal term for a variable, the thing you want an average of, the variable we usually place on the **x-axis** (the one that goes across the bottom of a graph).

The lower case **i** and **n** are your sample size. It goes from i=1 to i=n, with **n** being the number of data points (observations, or your sample size) that you have. It is the range you are summing over. And that **n** is repeated in the denominator.

Dow Jones Industrial Average

2011 distribution by 30 stock prices

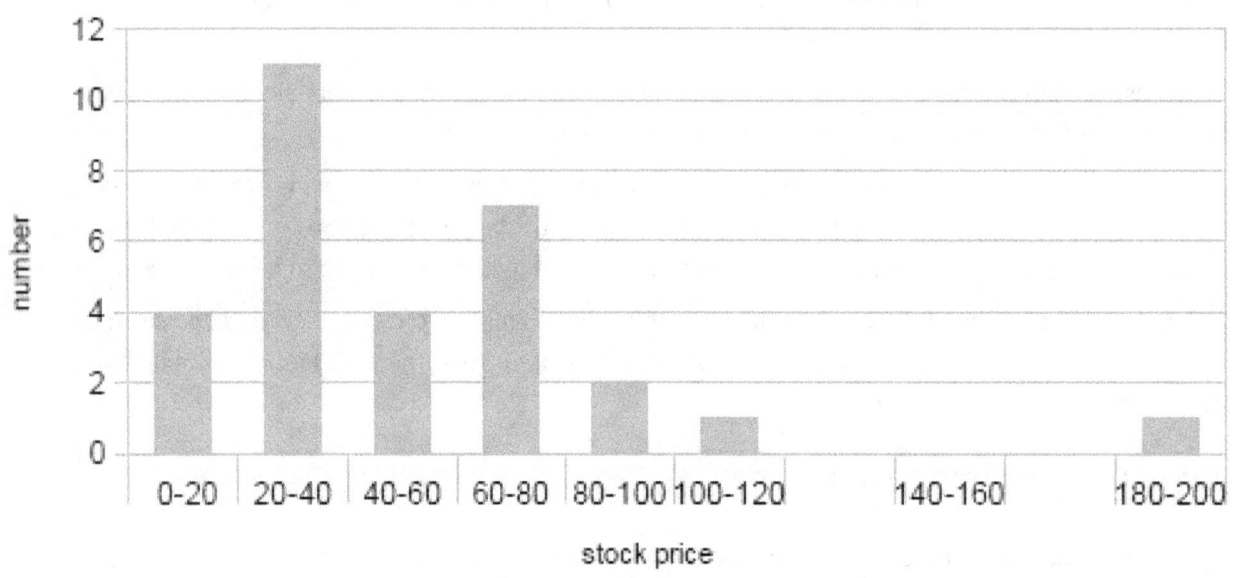

Example: Let's say that we want to calculate the Dow Jones Industrial Average (DJIA). There are 30 stocks in it, so 30 is our "n" and the stock prices are our "x," so we add up the 30 stock prices and divide by 30. The stock prices vary from $6 (our "i") to $187 (our "n"), and their average is $52.

$52? How can that be when the DJIA is 12,000? There must be something else going on in that formula, right? Actually, no, there isn't. The reason the DJIA is not 52 is mostly because of stock splits. That is, when a stock gets to be high in price, the company issuing the stock tends to split it—so that people who used to own 100 shares suddenly own 200 shares. (That's a simplification, but the principle is the same.) So now, instead of dividing by 30, the people who keep track of the average divide by .132129493. You can think of it as what the price today would be if you could have bought the stocks in the average 100 years ago.

The DJIA is truly that simple. Just a simple average we track every day. And it was started by a reporter who wanted to say whether the market was going up or down, and thought it would be easier just to look at one number. That a number which carries such importance in our economy is actually so simple displays the value, and predictive capability, of a simple average. Probably more predictive models have been attempted on the stock market than on anything else.

The calculation of an average is simple, but that does not mean that it gives a result that is descriptive. That leads into the next terms: normal distribution, and the related bell curve.

Normal Distribution and a Bell Curve

The reason the average (or arithmetic mean) is so important is that this measure is important both as a description and as a predictor. If the average is not a good description of the whole, its usefulness is limited and its predictive capabilities can be seriously impaired.

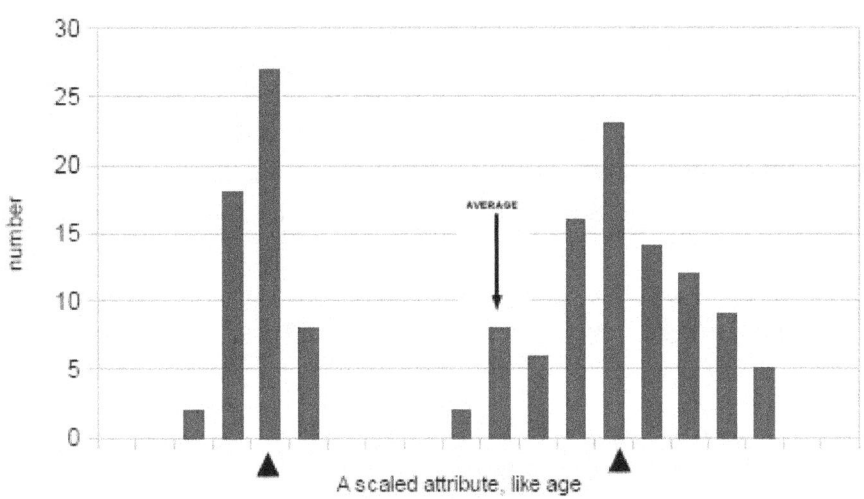

Example: This example portrays a hypothetical distribution example, say users of a consumer product by age range (a popular marketing tool). Let's say that you have a product that appeals to two distinct age ranges as in the two distinct sections of blue bars. If we were to calculate an average for the whole group, it would fall where the vertical line is, which obviously is not a good descriptive number. Instead, we need to realize that we need two numbers—an average for the left (younger) group and an average for the right (older) group, as marked by the black triangles. If you were planning an advertising campaign for this product, the average of the entire group would be calculated correctly, but it would not be useful.

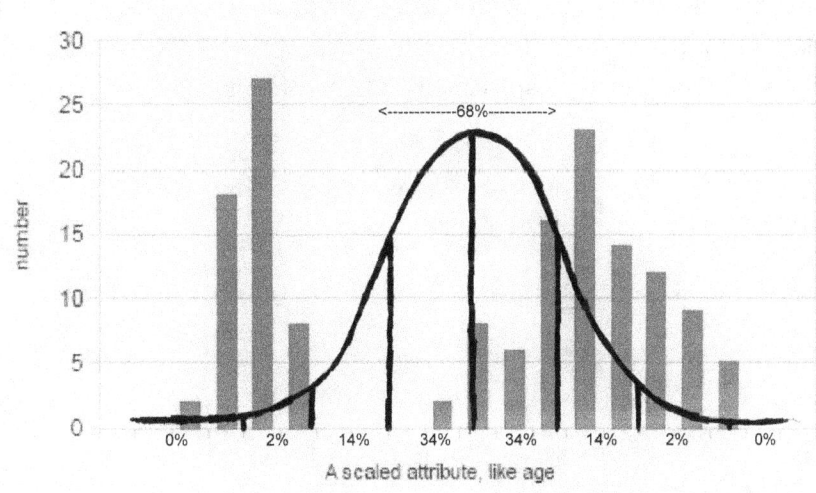

Of course statisticians have methods for making sure that the data is going to be accurately descriptive. This brings us to the concept of a **normal distribution** or **bell curve**, so named because the shape of the graphed distribution (the black line above) is in the shape of a bell. It is considered the most important probability distribution in statistics, and it is used as a model for even complex data (such as in error and uncertainty calculations).

Probability Density Function (Gaussian Function)

$$f(x) = \frac{1}{\sqrt{2\pi\sigma^2}} e^{-\frac{(x-\mu)^2}{2\sigma^2}}$$

The left side of this equation is called **f of x**, where **x** is our variable, just as it was before, and the *f* indicates that it is a function, meaning a relationship between two quantities.

The letter that looks a little like a *u* is the Greek symbol **mu**. It is the arithmetic mean, which is located at the peak of the bell curve—remember that "arithmetic mean" just means "average", the way most people think of an average.

The funny-looking o with the extra line is another Greek letter, a small sigma. It measures the **variance** (see next section), meaning the spread of the data across the graph.

You probably remember pi. It is just the standard constant pi (~3.14).

In a bell curve, by definition, over 2/3 of your data will fall within 1 standard deviation on either side of the arithmetic mean. From the graph above, it is clear that the blue bars do not come close to what an ideal bell curve distribution would have been. The analysis of the standard error would tell a statistician that the data does not fit. Care should therefore be used in the application of the arithmetic mean.

Variability Measurements: Variance and Standard Deviation

Variance is a measurement of how similar all your data are, especially the clusters of data around the mean, as in that bell-shaped curve ideal. If the data are spread far out (as in the 30 stocks in the Dow Jones Industrial Average) or clumped weirdly (as in the graphs above), the variance (and its closely-related **standard deviation**) will be high.

Variance examines the value of each piece of data, calculating how different it is from the average (arithmetic mean), and squaring it, then summing all those calculations together.

Variance formula:

$$\sigma^2 = \frac{\sum (X - \mu)^2}{N}$$

Variance is represented as the small **sigma-squared** on the left side.

Mu is the mean.

N is the number of data points (sample size).

X is our variable.

> **Example:** Let's do two simple examples just to see how much variance is affected by data that are spread out, compared to data that are clustered together.

> | |
> |---|
> | Data set 1: 1, 3, 5, 7, 9 |
> | Data set 2: 3, 4, 5, 6, 7 |

> For both, it is clear that **mu** will be 5, the average.

> The variance for data set 1 is:

> $(1-5)^2 + (3-5)^2 + (5-5)^2 + (7-5)^2 + (9-5)^2 = 16 + 4 + 0 + 4 + 16 = 40$

> The variance for data set 2 is:

> $(3-5)^2 + (4-5)^2 + (5-5)^2 + (6-5)^2 + (7-5)^2 = 4 + 1 + 0 + 1 + 4 = 10$

> As you can see, data set 1 has a bigger spread of the data (1-9) than data set 2 (3-7), and that is reflected in four times as much variance.

Standard deviation formula:

$$\sigma = \frac{\sum (X - \mu)^2}{N}$$

Assuming you have followed this far, standard deviation is very simple—really!

It is simply the square root of the variance. Whew! Check the formula for variance and you'll see that they are very similar.

For our two data set examples under variance, the standard deviation for set 1 is the square root of 40, or about 6.3; and the standard deviation for data set 2 is the square root of 10, or about 3.2. So you see that while the variance was four times as great for data set 1, the standard deviation is just twice as much.

Standard deviation is the most popular tool for measuring variance. It is the standard deviation measure that is used in the bell curve formula. By definition, 68% of the data values will fall within one standard deviation (above or below) the mean.

Standard deviation does not just express the variance. It is also used as a common measure of confidence in statistical conclusions. In fact, in polling data, it is the statistic used to measure the "margin of error," or random sampling error in the results.

These concepts of what an average is, or what a probability is, together with the confidence level associated with them, are important for any inference, such as a forecast of the future.

Forecasting

Statistical forecasting is one of the main types of inferential statistics used. The forecasting techniques use past data to predict future data by searching for a pattern or trend in the past data.

> **Simplest Forecast:** The simplest method is called a **naïve forecast**. It says the best prediction of the next value is simply the current value—simple indeed! It can be used by itself but more often is used as a comparison for more complex methods. It is a cheap and easy method, but is applicable only to trends that change slowly over time, called a **stable series**.

> This truly does sound naïve, too naïve, but it is should be pointed out that this is the model used for stock price evaluation. There still are no models that beat this, and certainly it is not for lack of trying! (Some of the models that have been tried include autoregressive integrated moving average, exponential smoothing, extrapolation, growth curve, and weighted moving average, among many others.) And obviously, if you are making a prediction from one value (like the current average price), that value had better be a good one.

This is the place where the fancy prediction formula with lots of Greek symbols belongs. The problem is that there is no single right forecasting method to use. There isn't even a standard on what type of data to use, including qualitative (subjective opinion/judgment) as well as quantitative (numerically-measured, such as averages) data.

The main assumption in forecasting is **exchangeability**, which means that the future is expected to behave like the past. While this tends to be true in the short-term future, it can break down in the long term.

One of the main ways to measure the value of a forecasting model involves running the model through several tests with other data, so that you can see how well it performs. For instance, if you have 100 data points, you might design the model on points 1-90 to predict 91, then 1-91 to predict 92, etc. Or you might examine a totally different but related sample (one where you would expect this model to work equally well) and see how it performs on both sets of data.

Forecasting is far from an exact science. If it were, your stockbroker's recommendations would always work! However, there are a lot of forecasting applications that do perform valuable functions, such as many of the applications used in supply chain management and transportation forecasting, in addition to product forecasting, political forecasting, sales forecasting, and many more.

Hypothesis Testing

One of the simplest inferential tests is used to answer a simple yes or no question, such as "Is this new drug effective?" and "Is the water quality good?" The data that are input are often probabilities or arithmetic means.

What you do in this test is set up a hypothesis that you attempt to disprove or prove. For instance, for a new drug test, the hypothesis would be that taking the new drug is no different from taking no drug.

There are five parts to hypothesis testing:

1. The truth is unknown

2. Make a "null hypothesis" (Ho) and an "alternative hypothesis" (Ha)

3. Identify and derive the test statistic, considering any statistical assumptions

4. Identify the rejection criteria, based on the distribution of the test statistic

5. Run the test, compute the test statistic, and conclude whether to reject the null hypothesis or to not reject the null hypothesis

The null hypothesis (Ho) is what you want to test, such as "Is this new drug effective?" The way that you phrase this, as a positive or a negative, affects the test results. The conclusion is not actually to prove or disprove the null hypothesis but to reject or not. That sounds so similar that it is hard to see the subtle difference; it is the difference between failing to reject and accepting, and they are slightly different.

- The much more common test is designed when Ho is worded opposite to what you want to prove, and it is called **reject-support**. As an example, in a test for a new drug, Ho is "there is no effect" for the new drug, and you are hoping to reject this with a high degree of confidence.

- In an **accept-support** design, the null hypothesis is what you want to prove (for example, taking the drug makes a difference). This sounds like an inconsequential difference, but it is actually very important in the confidence statistics, as will be seen below.

The test statistic is the measurement criterion used in the comparison of the two groups—for example, abatement of disease symptoms, level of quality, or level of probability. It is the crux of what you want to test.

The rejection criterion must be pre-specified for the test to be valid (usually 10%, 5% or 1%), and it is the criterion for when you accept Ho and when you reject it. The conclusion is the evaluation of whether the null hypothesis is rejected or accepted, and an examination of the confidence of this conclusion.

Even when you reach a conclusion with significance (that is, you accept or reject the null hypothesis), this conclusion could be wrong. This can happen in two ways.

- Type I error: Rejecting Ho (the null hypothesis) when it is true

- Type II error: Accepting Ho when it is false (that is, when Ha is true)

Another way to look at it is to use a table like the one below.

	Ho Is Really True	Ho Is Really False
Accept Ho	Correct Acceptance	Wrong-Type II-b (beta)
Reject Ho	Wrong-Type I-a (alpha)	Correct Rejection

It may seem that there is not much difference between Type I and Type II, but there actually is. For instance, in a court case, where the presumption is innocence (Ho), finding the defendant guilty when he is innocent (type I) is very different from finding him innocent when he is guilty (type II)!

> **Example:** The Coke versus Pepsi in the last section was actually designed as a hypothesis test. Ho was that your friend could not tell the difference, and Ha is the probability level of selection of Coke by random chance. In this particular example, the statistical test for confidence was intertwined in the measurement. Since it was measuring probability, that was the confidence level! You knew in advance, by looking at the probabilities, that you would reject it only if it was below 5% (actually less than or equal to 1.4%).

Like sample size, the formulas for statistical confidence depend on the model run. But there is more to be said about the calculations of these errors.

- **Type I error rate (false rejection)**—is specified at a certain level (usually at or below 5%), and this is the only error that many people are concerned with. It can be thought of as a "false positive" in a "reject-support" design, and it is a waste of effort and money. It is particularly difficult to refute, as tests often are not rerun.

- **Type II error rate (false acceptance)**—is also kept low, but generally not as low, often at a 20% or less level. The inverse of this error (1-b) is called the **power of the statistical test**, also called sensitivity. This measurement is a function of possible distributions under Ha.

And it should be noted that in essence, running an "accept-support" test reverses the Type I and Type II errors found in a "reject-support test." So test design is important.

Hypothesis testing can involve one sample (measured against a value, like our Coke example) or two samples (Ha is another group). Two sample tests can be paired (the groups are alike) or unpaired. Ha can either be two-sided or one-sided, called two-tailed or one-tailed in statistical jargon. This is a part of the test parameters, identified in advance: "Is any difference good enough, or must the difference be in only one direction?" For both the drug and quality examples, it would be one-tailed (only positive differences are acceptable).

There are two other terms important in hypothesis testing:

- **Control**—The data in a statistical hypothesis test can be from a controlled experiment or an observational study that is not controlled. A controlled experiment is one in which the two groups are not picked randomly but are deliberately controlled to try to keep them identical and each statistically "normal".

- **Blindness**—It is important that the people involved in the data groups (the test group and the control group) do not know which group they are in. When both the sample people (for example, patients) and the tester (for example, the doctor administering the medicine) do not know which is the real medicine and which is not, it is called a "double-blind" test.

Afterword

It should be noted that statistics results must often be interpreted, and even finding statistical significance does not necessarily mean that the result is significant in real-world terms. For example, the drug study may in fact show that drug has a positive effect but it may be so small an effect that the drug does not help substantially, especially to warrant its high price.

This section examined inference based on one variable—stock price, drug effectiveness, picking a coke. The next section will address a model that considers multiple variables simultaneously.

Advanced Inference

In this section we are going to examine two valuable mathematical inference methods: regression and Bayes' theorem. Both regression and Bayes' theorem are used for prediction and forecasting in many fields addressing problems applicable to everyday life. To some, they have an aura of mystery around them. However, the concepts are readily understandable.

Statistical Assumptions

In statistical inference, there are modeling assumptions and non-modeling assumptions. It is generally important to be aware of the assumptions so that they are addressed beforehand or tested afterwards, so that you make sure the predictions made are accurate and valid. Non-modeling assumptions include that the sample is random and the observations are representative of the problem being studied. Modeling assumptions include probability distributions, certain structural assumptions, and cross-variation.

Correlation vs. Causation

Statistics can only confirm correlation; it cannot confirm causality. This important point is often overlooked, even by experts.

Let's say that we can statistically prove that when one thing changes, another changes at the same time, in the same amount, and in either the same or opposite direction. This is what is meant by high correlation, and it is a valuable piece of information, as that relationship can then be exploited in practice. For instance, if you were to find that each time the price of oil went up, Exxon's stock price went up, then you could keep informed on oil price and quickly go out and either buy or sell Exxon stock each time oil prices changed. In statistics, this relationship is called **dependence**, and the extent of the correlation can be measured. **However, even perfect correlation is not causality.**

For instance, thousands of people looking to get rich have looked at the stock market to try to figure out how to predict future stock prices. They have looked at a lot of reasonable items (such as earnings, competition, and product demand) and a lot of unreasonable items (such as the length of women's skirts). Yes, it really was found that when the fashion called for women's hemlines to rise (the skirts became shorter), the stock market also rose, and when hemlines fell, so did the stock market. This is a correlation. But it does not imply causality—the hemlines falling did not cause the market to fall, nor did the market rising cause the hemlines to rise.

So how could we have correlation without one causing the other? What kind of relationship would that be? It is useful to think of it as if both were caused by something else.

For instance, perhaps the market went up when people were optimistic about the future—maybe wages were going up, home prices were going up, and unemployment was going down. In other words, everything in people's lives was going well, and they were happy. Maybe when they are happy, women like to wear shorter skirts. So it was happiness that caused both to go up—they were not reacting to each other but to something else entirely.

It is important to note that **there is no statistical test for causality.** All the statistical methods can measure is whether a relationship exists and how "strong" that relationship is (do they always react the same way in the same amount?). But no method exists for determining whether or not one causes the other to change, even though it is considered that hypothesis testing with an appropriate control can be a determination of causation.

Thinking that correlation proves causality is called a **logical fallacy**. It even has a fancy Latin name: *cum hoc ergo propter hoc* (translated as "with this therefore because of this"). Even scientists have been fooled into thinking that there was causality when there was not.

> **Example:** When hormone replacement therapy (HRT) became popular among women, scientists looked at the incidence of coronary heart disease (CHD) in women. There was a strong positive correlation that women talking HRT had a lower level of CHD, and the assumption was that the hormone medicine also worked for the heart disease. Only after further tests was it found that the women taking HRT had more money, ate a better diet, and exercised more. It was these other factors that explained the lower rate of heart disease, not the medicine.

The cautionary tale here is that even experts get this wrong, because it is so easy to fall into the trap. We see a strong relationship, and the causality just seems so intuitively obvious that we assume it to be the case. On the other hand, there may be a causality relationship. As Edward Tufte (a famous statistician) pointed out, "Correlation is not causation, but it sure is a hint."

Regression

Regression analysis has been used for over 200 years. In its most basic form, regression can be thought of simply as "line-fitting," although statisticians think this is a derogatory description. Regression is widely used in such diverse fields as accounting, engineering, environmental science, financial analysis, machine-learning, marketing, medicine, psychology, and sociology, among many others.

The forecasting done through regression is not a prediction of the future based on past data. Instead, regression is looking for relationships between variables. It looks to see what happens if you change one variable; how does it affect the other variable? It is predicting what one variable will be based on another variable.

A simple type of regression analysis is price and sales. Intuitively we know that as you raise the price of your product, you will have fewer sales. Regression analysis looks at issues just like this to say that if you changed your price today, this is how your sales would change. It is not a tool to forecast far in the future, in this sense.

In regression terminology, there is a **dependent variable** (like sales). This is the **predicted variable**, the variable that you want to forecast based on changes in other variables. It is also called the **y-variable** simply because the variables are often graphed, and by custom the dependent variable is plotted on the y-axis (on the left side pointing up, the vertical axis).

The variable that you want to examine for its impact on sales (like price) is called the **independent variable** or the **x-variable**, and it is plotted on the x-axis (on the bottom pointing right, the horizontal axis).

To start, you simply collect data, such as what your sales are at each price. There is no constraint that it be at specific price points; it is simply a list of what is available. You need an adequate sample size, but it does not need to be thousands or even hundreds. While under 15 will probably not be sufficient to detail the relationship with the confidence you would like, over 40 probably will. In statistics terminology, this data set is often called **observations**.

Next, you run the regression model to determine the relationship. It works exactly like plotting the data on a graph and drawing a line that will best fit through the data. The graph is called a scatter-plot, because the data points are usually scattered throughout. If the relationship is strong, the rough placement of the line should be quite obvious, like this:

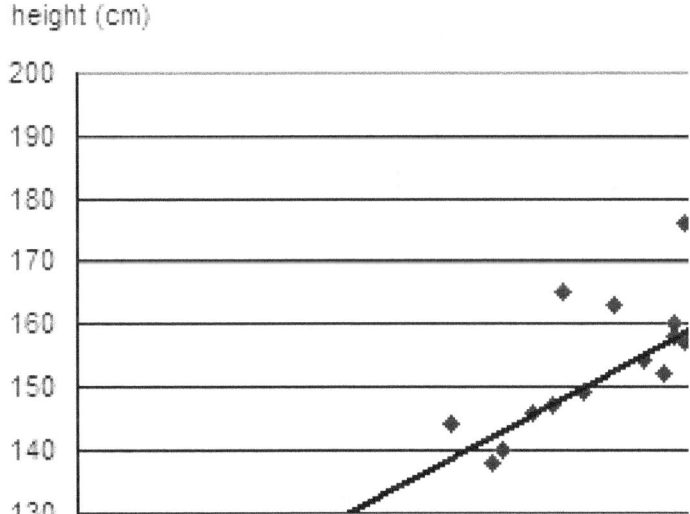

Yes, conceptually it really is that simple. It should make you wonder why regression seemed so mysterious and blackbox-like! Actually, this is exactly what any good statistic should look like. When all the data are graphed, a good statistic should make you say "Of course!" The line above is obviously a good fit to the data, displaying so easily the relationship between these two variables, with no calculation necessary. It should be just like when you look at a bell curve and can intuitively find the average. As you might expect, when the regression graph looks obvious, you will get a high reliability measurement in the analytical statistics—variance and standard deviation.

The graphed example is called a **linear regression**; a straight line is fit to the data, and the line is called the **regression line**. An alternate choice is **non-linear regression**, where the line can curve. Often both methods are tried to see which works better, and more than one independent variable (more x-variables) can be included as well.

The graphed example displays a positive relationship between the two variables—not in the sense of good or bad, just directionality. As the independent variable increases (going to the right on the x-axis), the dependent variable also increases (going up on the y-axis). There are also negative relationships where the line points down, which is the relationship you would expect in our sales-versus-price example; you would expect that as the product price rises, there will be less sales.

Linear regression formula: $y = a + bx$

y and **x** are our dependent and independent variables.

a is the "intercept point" where our fitted line crosses the y-axis.

b is the slope of the line (positive or negative).

While that looks fairly easy, the formulas in the mathematical models that are used to determine the best fit line do not! For the purposes of this book, we'll skip them.

Non-linear regression formula: $y = f(x1, x2, …, a0, a1, a2, …)$

This is not very helpful in understanding the relationship, but the non-linear line can take many different shapes, so the formula is a function, as indicated by the *f*.

Regression models aim to insure the best fit, since that will be the best predictor. The idea is to do this by calculating the line with the least variance. The most popular is the **least squares** approach, as seen in the last section. However, there are other approaches that will maximize the fit of the data according to another measurement, such as **least absolute deviation**.

Specifically, least squares defines the best line as the one where the observed y-values (the dependent variable, like sales) varies as little as possible from the line. The vertical distances from the observed values to their predicted value on the line are called **residuals**, and these residuals are referred to as the errors in predicting y. As in any prediction or estimation process, you want these errors to be as small as possible. Since the residuals are squared, the regression line is sensitive to the spread of the data, especially **outliers**, data points that are far off the regression line. The model will try all possibilities of lines to find the one that has the least squares value of error—that is, the smallest sum of the squared residuals, or the diagonal line where the spread of the data is the smallest above and below it.

There are more statistics that can be examined to analyze how good the fit of the line is, such as analyses of the pattern of residuals, hypothesis testing, F-test of the overall fit and t-tests of individual variables, and testing of the assumptions. These refinements are required if the sample size is small or if the errors are not normally distributed.

Due to extreme sensitivity to outliers, quality of input data is very important. A regression model will often be run several times excluding the more extreme input data, to make sure that there is not undue sensitivity to an instance of aberrant data.

Linear regression has many practical uses. Most of them fall into two broad areas: prediction and quantifying the strength of the relationship of various independent variables to the dependent one. When it is used for prediction or forecasting, the line will be able to give a value for y given any value of x, even when there is no information on that particular y in the data set. Many statisticians feel there is a distinct difference between predictions that are contained for y-values within the data set and those without—termed interpolation and extrapolation respectively; if you extrapolate the line past the data, you may think it is still straight, but in fact the line may curve.

When it is used to quantify the strength of variable, regression can be used to examine any x variable in relationship to the y variable, to see how strong the relationship is, or in fact if there is any relationship at all. It is also used to understand which among the independent variables are related to the dependent variable and to explore the forms of these relationships.

Remember, **correlation is not causation**. Regression analysis only determines a relationship and how strong it is; it generally cannot determine causality. As mentioned above, statistician Edward Tufte pointed out that "Correlation is not causation, but it sure is a hint." Typically, it will take other kinds of scientific testing to confirm the causality hinted at by regression.

For example, regression analysis first suggested that cigarette smoking caused lung cancer. Based on regression analysis, labels were added to cigarette packages warning that cigarette smoking *might* be harmful to your health. It was only after further tests confirmed the causality of the relationship that the label was changed to its current version.

Bayes' Theorem

Although Bayes' Theorem has been around for over 250 years, it has become very popular in recent years as an important tool in artificial intelligence (AI) and many other diverse fields, from philosophy to forensic science. Bayes' Theorem was even used by British philosopher Richard Price to try to prove the existence of God in 1763! It is an extremely useful tool in estimating probabilities, given knowledge of certain related probabilities.

It is useful because it can address many variables and many probabilities simultaneously. Of course, that means that Bayes' Theorem models are not easily graphed, so their answers cannot be easily visualized. And that is one of the more interesting parts of Bayes' Theorem results—they are often very unexpected, even counter-intuitive. Sorting out the impact of a lot of multiple probabilities is hard without a model. The Bayes' Theorem model can be considered dynamic, as it helps us to understand what we know, given the evidence we have and other information we discover, and how to do this in a quantifiable manner.

> **Example:** This is a real-life probability problem. Let's say you have a medical test—for example, a cancer test. The test comes back positive, which to you is very negative, as it means that the test says you have cancer. This is not a biopsy-type test that confirms cancer; it is a diagnostic test that has a high degree of accuracy but is *not* perfect. In other words, this test can yield **false positives**. That means that the test comes back positive when the correct diagnosis is really negative, which the biopsy will confirm. When people actually do have cancer, the test comes back positive 80% of the time. So—how much should you freak out?
>
> We know that the following probabilities have been confirmed as accurate.
>
> At age 40, 1% of women have breast cancer and 99% do not (the prior probability).
>
> A mammogram will be positive 80% of the time for women who do have breast cancer, and a mammogram will be positive 10% of the time for women who do not have breast cancer (the conditional probability).
>
> The question is, when someone gets a positive result on the test for breast cancer, how likely is it

that she has breast cancer? Most people, including 15% of doctors, do not get the right answer, thinking it must be close to 80%. In fact, it is quite different–a surprisingly low 7%.

The reason for this is that those conditional probabilities obscure the final probabilities. If you reword the problem, you can look at the numbers quite differently:

Out of 10,000 women:

100 will have cancer
  80 of them will get a positive mammogram (correctly) and
  20 a negative one (falsely)

9,900 will not have breast cancer
  990 will get a positive mammogram (falsely)
8,910 will get a negative one (correctly)

The total number of positive mammograms is 990 + 80 = 1070

The probability of a positive mammogram test for everyone is 1070/10000 = 11%

Of the 1070 positive mammograms, 80 are for women with cancer, so 80/1070 is 7%.

The mammogram test increased the probability of having cancer from 1% (before the test) to 7%. A significant increase, to be sure, over seven times the level as before, but 7% is far from 80%.

So the answer to the initial question is: Don't freak out until after the biopsy.

The surprising results in this example would be duplicated in any test situation where the percentage of people who have the disease is very small, such as the 1% here. This can also be applied to symptoms and how serious they might be. A symptom is virtually meaningless if as many as 10% of healthy people exhibit the symptom and the disease is relatively rare. Thus the accuracy of tests for rare conditions must be very high in order to produce reliable results from a single test, due to the possibility of false positives.

"Bayes Decision Tree"—Cancer Test

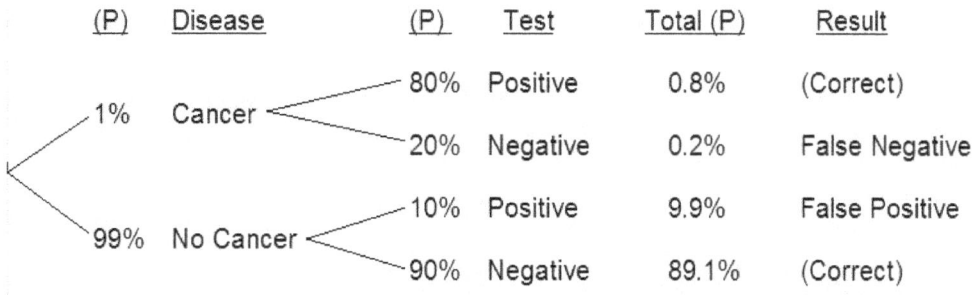

(P)	Disease	(P)	Test	Total (P)	Result
1%	Cancer	80%	Positive	0.8%	(Correct)
		20%	Negative	0.2%	False Negative
99%	No Cancer	10%	Positive	9.9%	False Positive
		90%	Negative	89.1%	(Correct)

These results could have been derived through Bayes' Theorem, where the mathematical formula is written as follows:

$$P(A\backslash B) = \frac{[P(B\backslash A)]\ [P(A)]}{P(B)}$$

where P(A\B) is the conditional probability of A given B

This can be thought of in English as: what we know, given the evidence = what we knew even without the evidence, adjusted by how good that evidence is.

Since this is still confusing, let's look at our cancer test example.

P(A\B) is the probability of having cancer (A) given a positive test (B)—that is what we want to determine.
P(B\A) is the chance of a positive test given that you have cancer (80%).
P(A) is the probability of cancer (1%).
P(B) is the probability of a positive test (11%).

So, to put it all together,

$$P(A\backslash B) = \frac{.8 \times .01}{.11}$$

Which becomes   **.07** or **7%**

Baysean logic is something we do every day: adjusting our probability calculation with new evidence. When a child says he didn't eat the cookie, but his mom sees cookie crumbs on his clothes, her estimate of his guilt goes up. These problems with conditional probabilities are complicated and not intuitive.

Conclusion

Statistics are useful to describe data and data relationships, especially when there is a lot of data or the relationships are complex. Statistics are invaluable in forecasting as well, and are used in almost every field to address and solve everyday problems.

But there is a large caveat. Statistical formulas must be applied carefully to make sure that the data are accurate and the statistics derived from them are a good fit for that data. Many statistics that are readily available in television and newspapers and on the internet have not been statistically evaluated for their validity.